THE COPY CATALOGUE

by Barry Biggles

Pantheon Books: New York

CONTENTS

It's high time the office photocopier came out of the shadows. This masterpiece of technology deserves more in life than churning out endless memos. Turn it loose on memorable **MESSAGES** (Section I). If you need to tell someone where to go, including which way up, try **OFFICE LINES** (Section II). If you need a hand to give you a pointer, or vice versa, look in **DIRECTIONS** (Section III). You may already have discovered the recreational aspects of the copying machine, but with THE COPY CATALOGUE you're set to make the most of them by generating items that are useful, attractive and amusing.

There are letters and numbers to suit all occasions in **CHARACTERS** (Section IV); you can combine initials and figures for personalized birthday and anniversary greetings. The price of commercial cards and tags these days just about takes the pleasure out of buying them, and few are even passably original. Yet the cost of custom-printing your own designs can also be discouraging. The beauty of THE COPY CATALOGUE is that you combine speedy production with economy — and still have the chance to add your own individual flourish. There are countless domestic and decorative functions for **LABELS** (Section V), so take your clever ideas and make them stick.

INVITATIONS (Section VI) presents a host of offers no-one could refuse. Make the most of informal occasions by sharpening your wit before the event. A real invitation conveys an air of excitement so much better than a phone call. The party mood continues in **CELEBRATIONS** (SECTION VII), spilling over with banners, streamers, paper chains and winners' rosettes. For the finishing touch, look in **DECORATIONS** (Section VIII) and consider the opportunities for dressing up or recycling commercial containers, household items or stationery. Here is your incentive to embellish with relish.

THE COPY CATALOGUE supplies professional designers' tips for making, mounting and colouring your creations. See what unusual results you can get by experimenting with moving images or 3-D objects. Feel free to overlay designs and words from different parts of THE COPY CATALOGUE and/or your own sources, and see what a surprising range of expression is possible with that familiar office fixture.

Anyone who lacks access to a photocopier at work can find a suitable machine in the public library or at art and stationery supply stores, and those who would simply like models on which to base their own hand-drawn work will find THE COPY CATALOGUE an invaluable reference. However you choose to use THE COPY CATALOGUE, you may regard the photocopier as your own designer and printer from now on. So go ahead; reproduction can be fun.

THE COPY CATALOGUE

Edited and designed by
Marshall Editions Ltd
71 Eccleston Square
London SW1V 1PJ

Copyright ©1981   by Marshall Editions Ltd

All rights reserved under International and Pan-American Copyright Conventions. Published in the United States by Pantheon Books, a division of Random House, Inc., New York.
Library of Congress Cataloging in Publication Data

Biggles, Barry.
      The copy catalogue.

      1.   Electrostatic printing—Amateurs' manuals.
2.   Printing—Specimens—Catalogs.   I.   Title.
TR1042.B53   686.4′4   81-47205
ISBN 0-394-74900-6   AACR2
Manufactured in the United States of America

Text and display setting Mayday London

98765

Very few materials are needed to transform plain photocopies into attractive finished items. Most of what it takes to cut, mount and decorate is already on hand in homes and offices. Here's a checklist: a ruler or straightedge and set square or right angle for clean lines and perfect corners; scissors and scalpel for cutting; rubber cement or clear paste and double-sided tape for mounting; an assortment of felt-tip pens and coloured paper for brightening up your subjects. Some tips: always keep your fingers well back from the cutting edge when using a knife, and use a sheet of heavy card or mounting board to protect the work surface. Keep white typewriter correcting fluid on hand to retouch unwanted lines. When using double-sided tape for mounting, get the position right first time as it is permanent. Use paste in moderation so as to avoid lumpy areas. If working with felt-tip pens that bleed, keep well within the outline of the subject. Remember you can use the back of your copy paper to write messages or affix other designs. The latest copiers offer increasingly more options. Some can do both enlarging and reduction of images. They can also copy on to coloured stock or gum-backed paper, which makes the production of labels, for instance, extra simple.

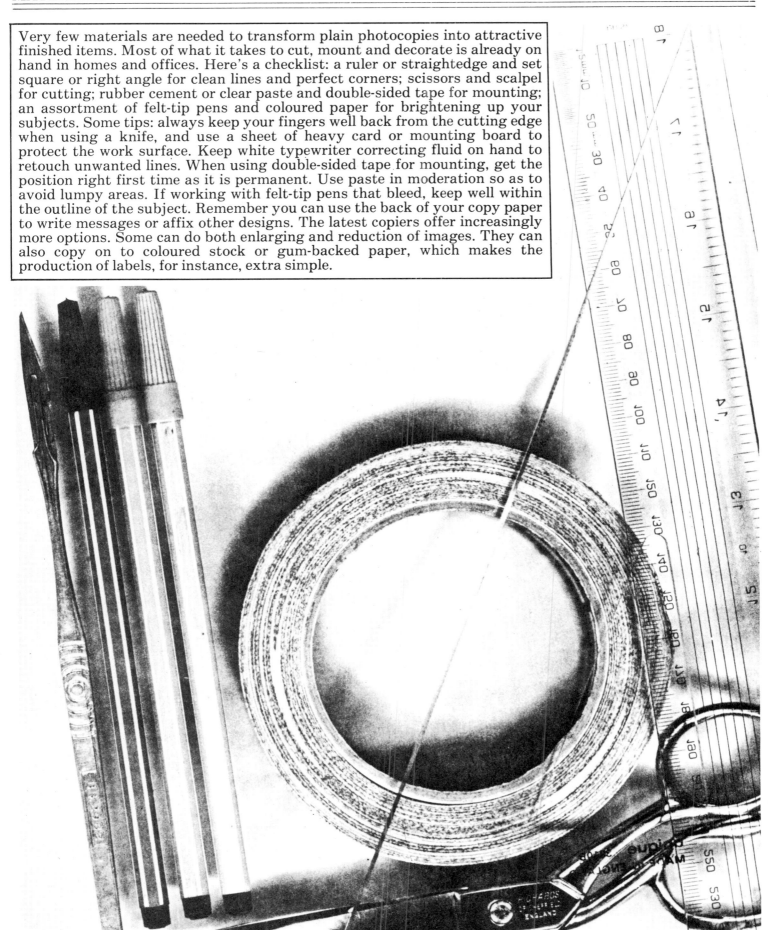

Choose your method of mounting a subject according to its end use — whether for office, display or mailing — and select the appropriate materials. Take care not to rush the job; some patience now pays off in quality. Think out the steps and then proceed. Some tips: don't use double-sided tape to mount a sign on wallpaper; it is so permanent it could lift the paper when you try to take the item down. Reserve it for plastic, wood, metal or gloss paint surfaces. As some glues are inflammable, keep naked flames away from the work area.

FRAMING. Two ways of mounting a subject you intend to frame.
1. Image centred on tinted card. Measure image area and add to this a mount area of which the top and sides are equal — say 3 inches — and the bottom slightly deeper — say 3½ inches — as this helps set off the image. Cut card to size and position image by marking its corner points on the card in pencil. Apply glue to image rather than card and fix in place, top corners first.

2. Window mounts. These are very attractive if you are using coloured card as you get a nice white edge showing from the inside of the cut board. First work out the area of image that is to show; then draw the window on to the card. Next work out your border areas, cutting out the window before the perimeter. Keep your fingers away from the edge of the ruler when using a knife.

DISPLAY SIGNS. Add the impact of a third dimension to messages like Back Soon, Hands Off, The Boss and so on. Cut out the image, working close to the line. Next cut out backing card which is at least the width of the subject and three times the depth, plus an extra half inch for the return flap. Paste the subject on to the middle panel, then fold along the top and bottom edges as shown RIGHT. Fold the return flap and fix with double-sided tape. Now your sign stands up and can be read from a distance.

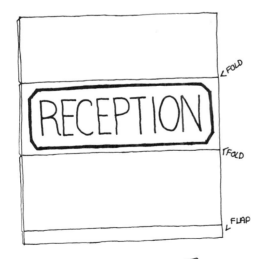

To make certain a subject stays flat, lay a piece of paper above it once you have positioned it; then lay a ruler along the edge of the subject. Draw the ruler towards you, pressing down as you do so. The same method may be used when you cover a subject with clear adhesive film. If covering a large area, have someone hold the film away from the subject as you draw the ruler over it. You may also use a cloth, smoothing to left and right from the middle outward. Should any bubbles appear, prick them with a pin and rub them down with your fingernail.

Most types of coloured felt-tip pens are suitable for decorating the images when copied. Spirit-filled pens tend to bleed, so if you are concerned with a neat appearance, try the pens out on scrap copy paper first, working up to an outline, so you can judge how close to the line you should work on your finished print. For many jobs, colour needn't fit precisely; it can simply be washed in. If using water colours, don't let the paint get too thin or the paper will stretch.

Coloured papers can be used as backgrounds to the images or cutouts can be made to offset them. Also, copies may be made directly on to tinted paper, provided the stock is the right weight and size for the photocopier.

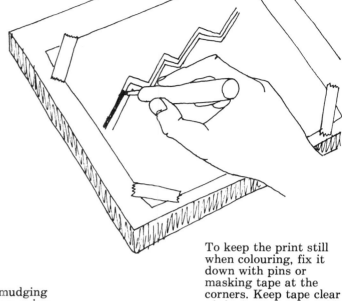

To keep the print still when colouring, fix it down with pins or masking tape at the corners. Keep tape clear of the image as it could lift it, or damage the paper.

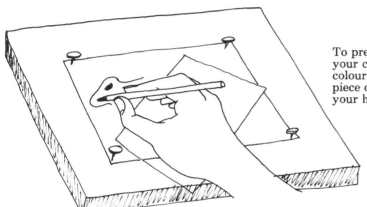

To prevent smudging your copy when using coloured pencils, lay a piece of paper under your hand.

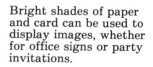

Bright shades of paper and card can be used to display images, whether for office signs or party invitations.

To prevent the image stretching when using water-based colours first wet the copy on both sides under running water without letting it get too soft. Lay the copy on a board and remove surface water with a sponge or paper towel, then fix down with tape — ideally the gummed brown paper kind you have to wet. Allow the print to dry before applying paint. When the paint has fully dried, cut the print from the board and trim to size.

If you wish to make a copy from a book and the image you want is in the middle of other pictures or artwork on the page, a mask will allow you to make a clean copy. When copying from a book do not press down too hard or you could damage the binding. Colour pictures reproduce comparatively dark so you may want to set the tone scale lighter before you start.

Draw out the mask area around the image on tracing paper and transfer to mask paper by laying the tracing on top, pricking through the four corners to mark the area to be cut out. If you prefer, simply measure the area around the image and draw it directly on to the mask paper.

You can mask an image directly on the photocopier's glass screen by laying down offcuts of paper — better still, two L-shaped masks — which can be adjusted to make a window around the image to be copied.

Use a board when cutting so that you don't damage the work surface.

Sometimes a shadow line appears when an image has been pasted into position and then copied. You can lose this by painting it out with white typewriter correcting fluid or designers' white paint. When you recopy your copy, the image should be clean.

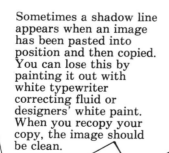

When the mask is made, use masking tape to fix it in position over the image to be copied. Avoid pressing the tape down too hard or you might damage the page.

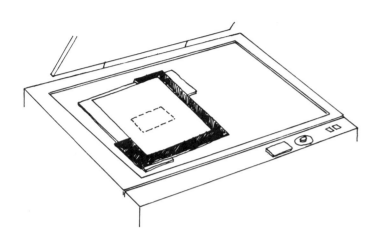

Copying objects as opposed to flat images can produce truly surprising and satisfying results. A still life of vegetables may create the background for a written menu; watches or jewellery might document insurance records; a baby's knitted shoe would suit a first birthday reminder. It can also be a drawing aid, translating into two dimensions articles difficult to sketch from life. Projects may be set up for children, such as arranging wildflowers, leaves and grasses collected on walks to make a print for hand-tinting and framing. Any specially pleasing motif could be duplicated repeatedly and used to give new life to, say, an old wastebasket. If you can't resist copying your face, do at least take the precaution of closing your eyes. Experiment with the tone control for varied effects.

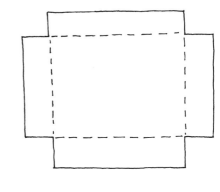

TO MAKE A DARK BOX. A dark box is needed to copy objects too tall for the machine lid to cover without crushing them. Measure the light table area, adding the desired height of the box to each side. Cut as shown. RIGHT. Fold sides down and tape corners.

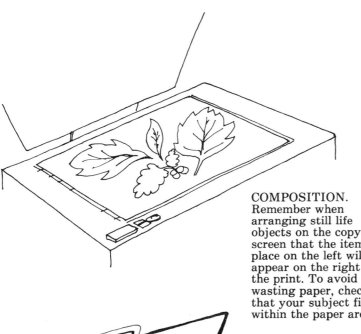

COMPOSITION. Remember when arranging still life objects on the copy screen that the item you place on the left will appear on the right of the print. To avoid wasting paper, check that your subject fits within the paper area.

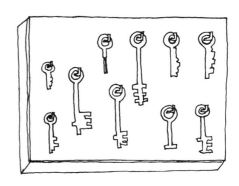

KEY BOARDS. Make a pattern with the keys all the same way up on the glass copy screen. You can get a black background by leaving the machine lid open. To get a really perfect black background lay a piece of black or dark-coloured card over the subject before making the copy. Mount the copy on a wooden board, protect with adhesive film and fix nails so that the keys hang over their copied images. As well as encouraging tidiness and routine, the board records the shape of any missing key.

The art of moving objects on the copying screen is by nature experimental. The knack is in moving the object in time with the light tube. The image will stretch from left to right, not top to bottom. Better results are achieved with subjects comparatively light in colour.

You can achieve some really striking effects that are both expressive in their own right and in combination with messages. The high-speed blur of a watch says time flies; a floating mug invites a morning-after reviver; the distortion-twinned image of a figure can say you're beside yourself, or that it's party time.

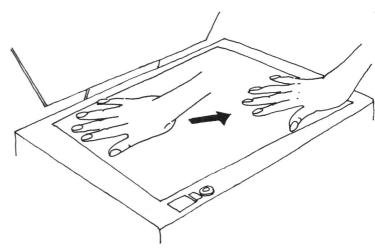

If copying your hand, for instance, try not to move it too soon or you will lose the finger tips. Moving your hand over the screen with the light tube gives the sort of result shown LEFT. Remember, the image will be reversed.

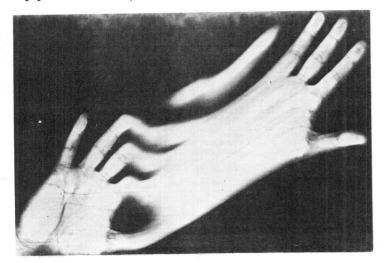

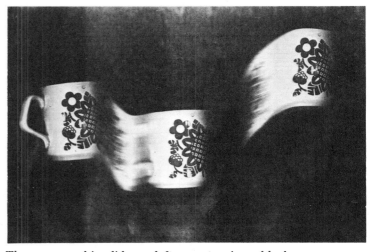

The copy machine lid was left open to give a black background.

The scissors ABOVE were fixed from underneath with tape to a white card, then moved across the light table. The same was done with the watch BELOW.

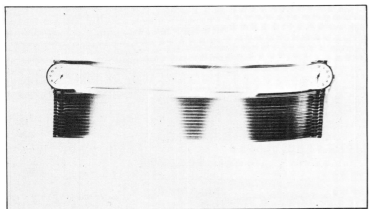

The test card is your first step to a perfect print. Use it to save both time and paper. With it you can check for a perfect background white; a clean, solid black line; a large area of solid black; and a tone area which reproduces without filling in.

Work out the requirements of the print you want to make with the help of the illustration below. The glass screen must be clean and the toner filled up in order for the test card to work.

If what should be a perfect outline is breaking up on your test print, check to see that the tone control is set dark enough and that the toner itself is not running out.

Use the tone control knob to achieve a perfect background white.

Failure to get solid black over a large area could be due to either insufficient toner in the machine or an improperly set tone control.

Check this tone area before reproducing from a black and white or colour print in a book. If the copy is still grey after adjusting the tone control, check the tone filter.

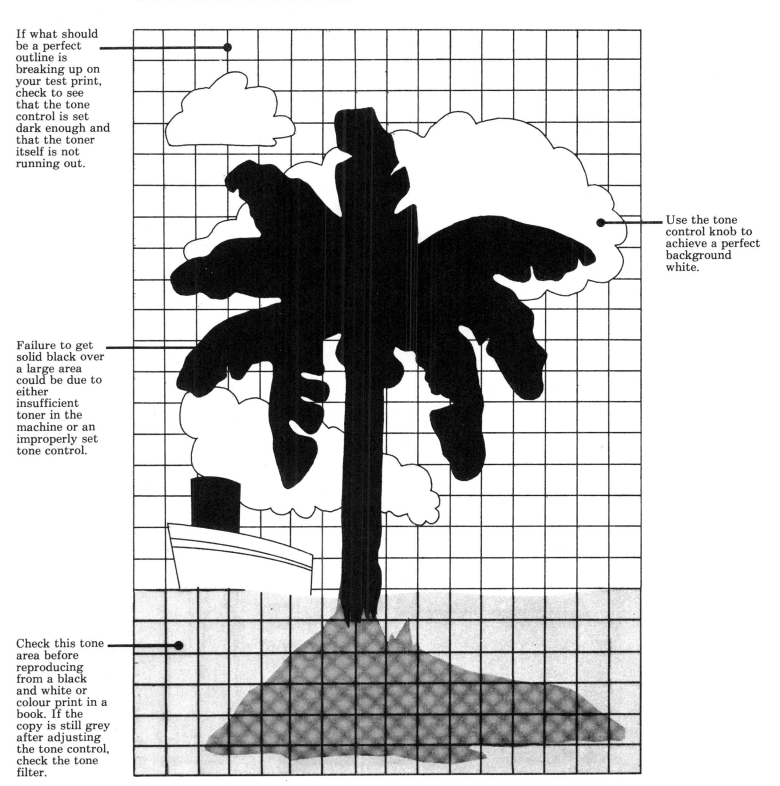

Auto motive. When you put the car in for servicing or repairs, a proper check list for the mechanic means your instructions should not be missed. The written record can also be a useful back-up in case of a misunderstanding.

Tell a phone. If you attach your message directly to the instrument in question you're in with a better chance of a swift reply than those who have simply added to the nest of paperwork on the desk.

A bright idea. If you have trouble getting your own attention, turn yourself on by sticking a message in a strategic place.

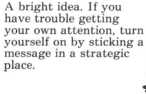

Snooper-proof. If you fold this message sheet in half and back on itself, the name and telephone number of the sender will be prominently displayed, while the message is discreetly concealed on the underside.

Say it again. You give any message more power by repeating it, and the pattern you make will have impact of its own.

MESSAGE

NAME/ADDRESS _____

_____

TELEPHONE _____

MESSAGE

NAME/ADDRESS _____

_____

TELEPHONE _____

MESSAGE

NAME/ADDRESS _____

_____

TELEPHONE _____

MESSAGE

NAME/ADDRESS _____

_____

TELEPHONE _____

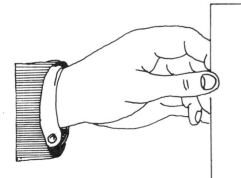

While you were out...

TELEPHONE MESSAGE FOR: _____

FROM _____ TIME _____

MESSAGE TAKEN BY _____ DATE _____

While you were out...

TELEPHONE MESSAGE FOR: _____

FROM _____ TIME _____

MESSAGE TAKEN BY _____ DATE _____

While you were out...

TELEPHONE MESSAGE FOR: _____

FROM _____ TIME _____

MESSAGE TAKEN BY _____ DATE _____

While you were out...

TELEPHONE MESSAGE FOR: _____

FROM _____ TIME _____

MESSAGE TAKEN BY _____ DATE _____

While you were out...

TELEPHONE MESSAGE FOR: _____

FROM _____ TIME _____

MESSAGE TAKEN BY _____ DATE _____

The last word from...........

So there!

THANK YOU THANK YOU THANK YOU THANK YOU THANK YOU

You were never lovelier

Hey sport!

FOLD

FOLD

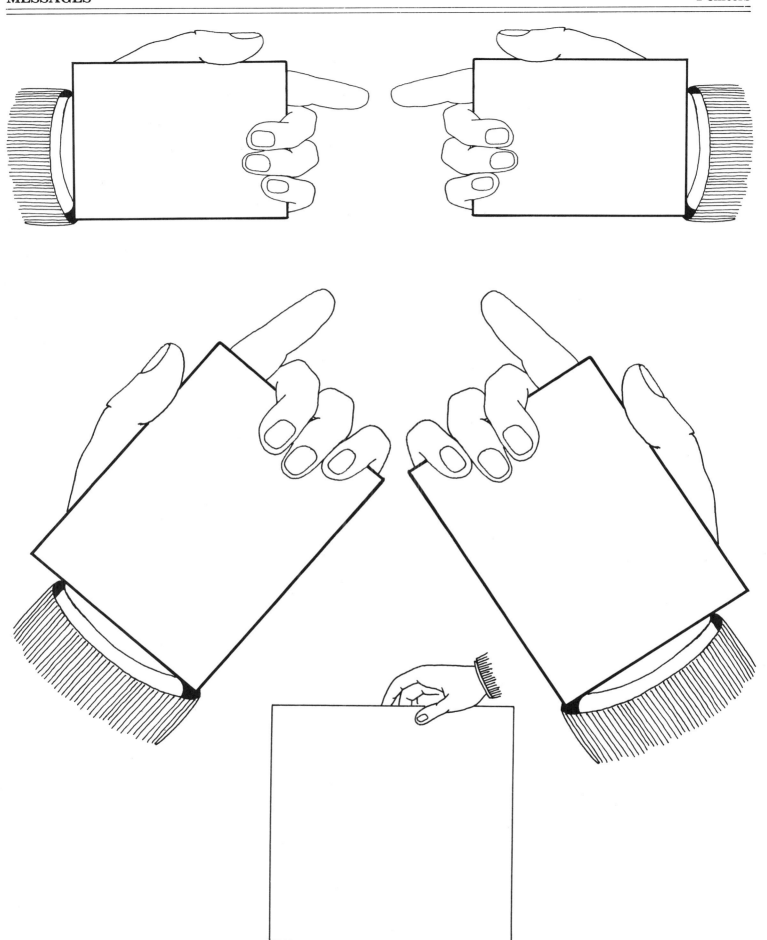

FOLD

FOLD

Get Well Soon

Get Well Soon

Get Well Soon

GET WELL SOON

Get Well Soon

Dear Baby-sitter
He/she/they/it will expect....

If you need help, call me/us at

Or phone   We shall be home by................

Dear Mr Mechanic....
Please fix

If it's all more complicated than I think, I can be reached at..........

....signed..................

YOU'RE MAGNIFICENT

You're Terrific

YOU'RE ADORABLE

You're The Greatest

LOVE YOU

TAKE CARE

SORRY!

HAVE FUN

THANKS

WELCOME BACK

HAVING FUN

HAVE FUN

MISS YOU

SORRY!

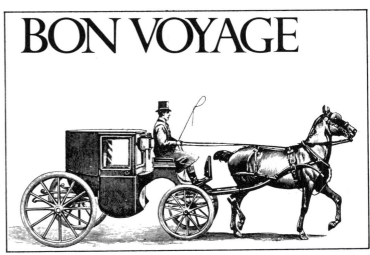

KEEP YOUR PECKER UP

STOP DOGGING ME

you are a dear

Don't Get Cocky

you old bore

I FEEL... sheepish

You will find the OFFICE LINES self-explanatory. Here are a few ideas for putting them to work. Several are suitable for home use too. Then there are those destined to·lead an amusing double life out of context. Don't hesitate to colour them in, and remember to put a protective finish — either varnish or clear plastic adhesive film — over signs intended to be permanent. Parcel stickers may be backed with double-sided tape. Even if these don't make order out of chaos, they'll make chaos better looking.

OFFICE

STORE ROOM

MAINTENANCE

Telephone

ACCOUNTS

RECEPTION

RECEPTION

REPORT HERE

ENQUIRIES

Deliveries

Do Not Disturb

Do not disturb

Knock and Enter

MEETING IN PROGRESS

KNOCK AND ENTER

KEEP CLEAR

NO SMOKING

NO SMOKING

KEEP SHUT

QUIET

QUIET

FERMEZ LA PORTE

KEEP OUT

IN OUT

IN IN IN OUT

OUT OUT IN IN IN

IN IN OUT

IN IN OUT

IN IN IN OUT OUT OUT

ENTRANCE

EXIT

WAY IN

WAY OUT

WAY IN
WAY OUT

ENTRANCE ENTRANCE

EXIT

EXIT
EXIT

EXIT

PUSH HARD

PUSH GENTLY

PUSH HARD

PULL HARD

PUSH OPEN

PULL OPEN

OPEN GENTLY

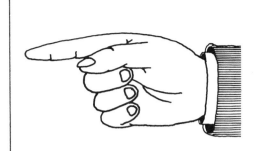

FIRE EXIT

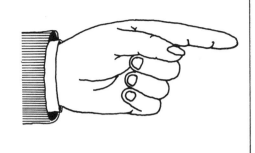

LADIES

Powder Room

THIS WAY

MEN'S

REST ROOM

EXPRESS
EXPRESS

EXPRESS
EXPRESS
EXPRESS

EXPRESS
EXPRESS
EXPRESS

URGENT!

URGENT!

RUSH

RUSH

PANIC!

◁EXPRESS▷

EXPRESS

DANGER

ACHTUNG

DANGER

DANGER

DANGER

CAREFUL! HAZARD

CAREFUL!

HAZARD

CAREFUL!

DO NOT TOUCH

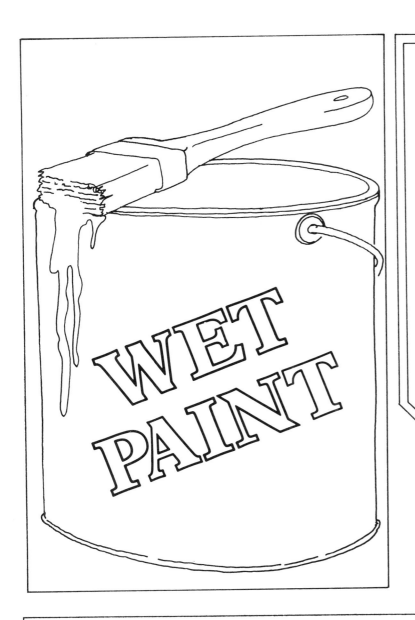

DO NOT TOUCH

WET PAINT

WET PAINT

NO FOOD

NO DRINKS

NO CHILDREN

NO DOGS

NO PUNKS

NO GROWN-UPS

NO BAREFEET

NO ROLLER SKATES

NO RADIOS

NO TOPLESS SWIMSUITS

HANDLE WITH CARE

HANDLE WITH CARE

HANDLE WITH CARE

Fragile

TAKE CARE

Fragile!

1.
2.DUMP

FROM THE BOSS

NAG LIST

1
2
3
4
5
6
7
8
9
10

COMMUNIQUE

PANIC!

THIS IS A SECRET

THE LAST WORD FROM

THE END

FOR YOUR EYES ONLY

A PARTY

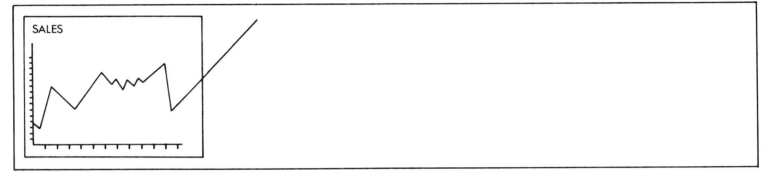

SALES

PLANNING

FINANCE

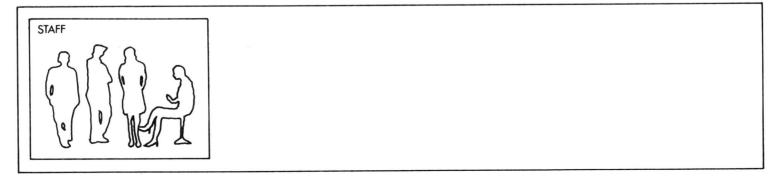

STAFF

31	30	29	28	27	26	25	24	23	22	21	20	19	18	17	16	15	14	13	12	11	10	9	8	7	6	5	4	3	2	1	

	8.30	9.00	9.30	10.00	10.30	11.00	11.30	12.00	12.30	1.00	1.30	2.00	2.30	3.00	3.30	4.00	4.30	5.00	5.30	6.00	6.30	7.00	7.30	8.00
Monday																								
Tuesday																								
Wednesday																								
Thursday																								
Friday																								
Saturday																								
Sunday																								

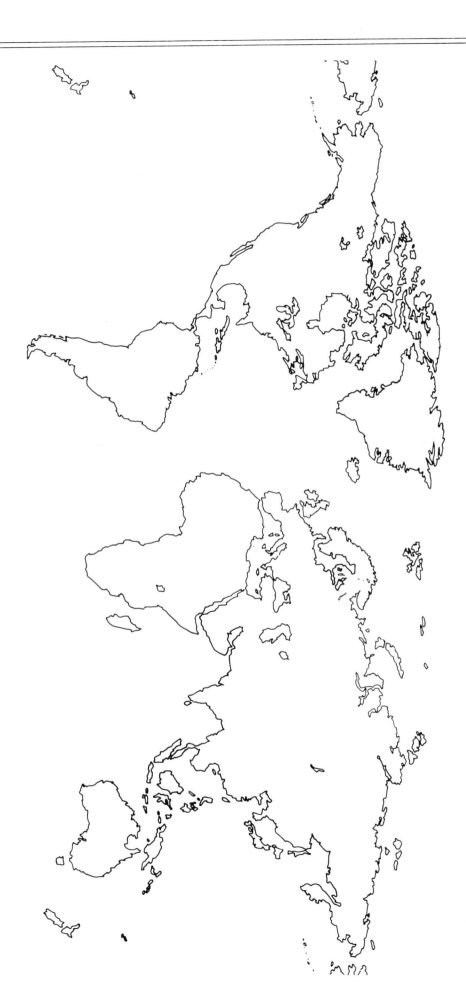

Hands up. The many hands on the following pages make light work of messages and invitations. Put them to dynamic use by cutting out the photocopied hand close to the outline; next cut out a piece of card the width of the hand and somewhat longer so as to project beyond the finger tips. Tape the back as shown. Use the card for your message. You can make a two-handed display by doubling up this technique, but using a single background card. (Be sure the card is long enough.)

Point the way. Attach one of the directional hands shown to the appropriate copy from OFFICE LINES. Paste down on backing card and trim to size.

Every indication. Make a handsome card by duplicating your chosen arrow pattern. Colour the arrows in, then cut back to the outline. Section off the card as shown RIGHT, using a soft pencil. Once the arrows are fixed into position, you can erase any visible pencil marks. Just the card for telling someone you may not know where you're going, but your aim is true.

Graphic art. Neat little arrows are always an asset on wall charts and graphs. May your own point ever upward.

Four-point plan. For this card, measure the area of a quarter section and quadruple it for the back. Paste arrows into position, then fold and cut as shown. It's bound to attract attention, so save it for when you really need to make a point.

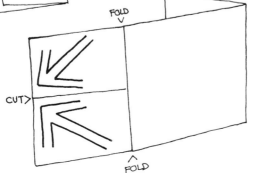

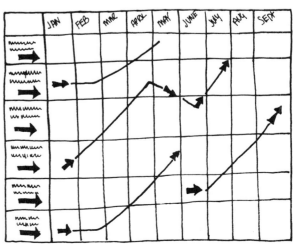

THIS WAY UP

THIS WAY UP

THIS WAY UP

THIS WAY UP

THIS WAY UP

THIS WAY UP

THIS WAY UP

KEEP LEFT

TURN LEFT

KEEP RIGHT

TURN RIGHT

UP

DOWN

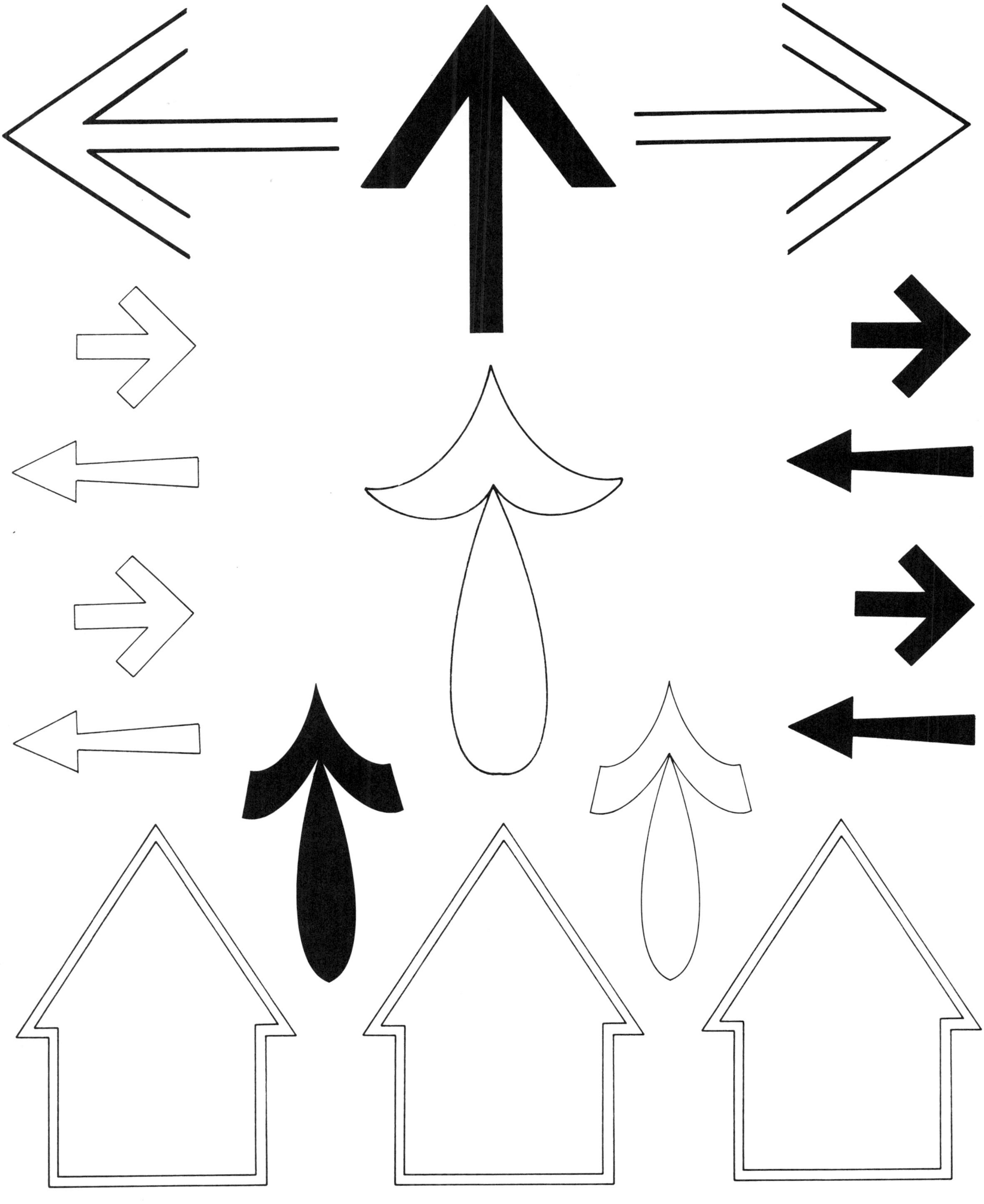

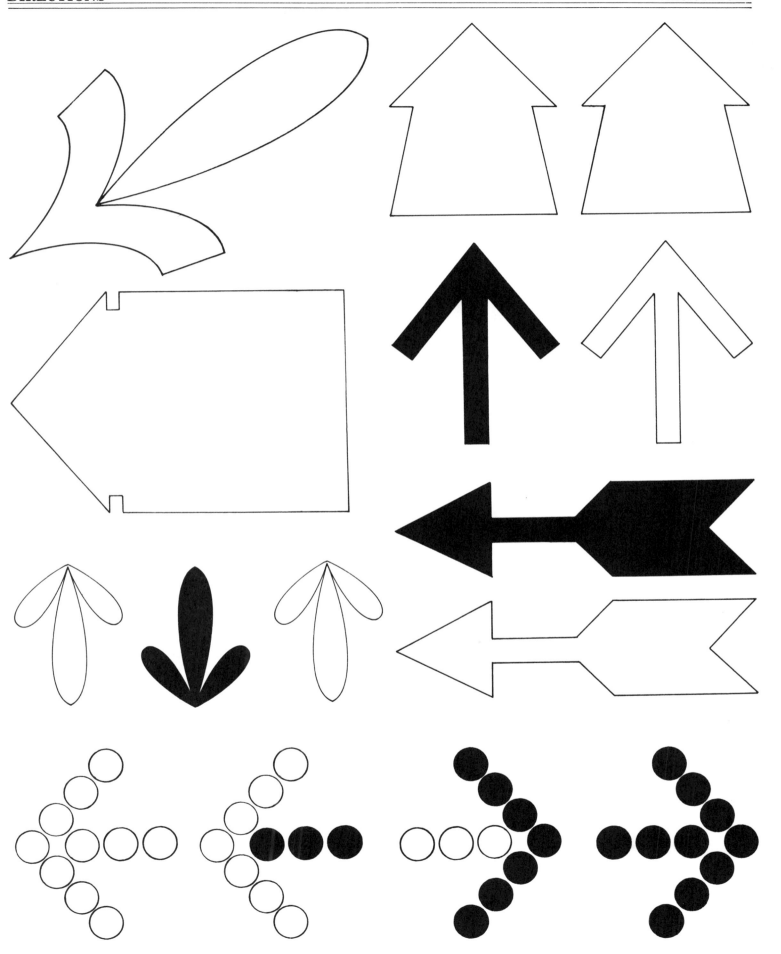

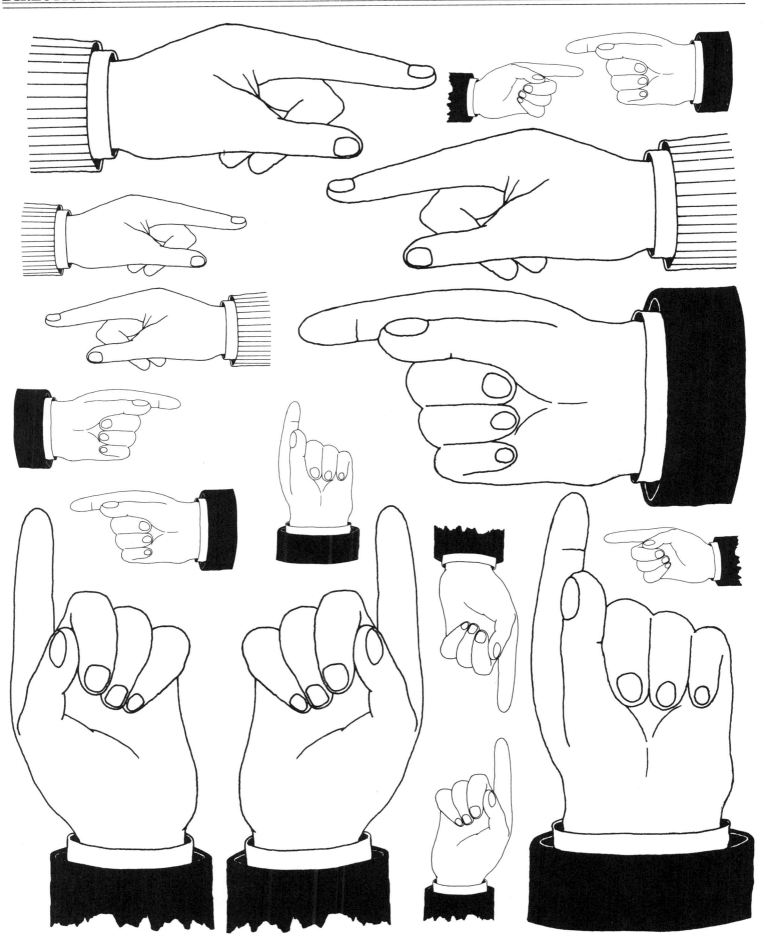

Got your number. Give the personal touch to anniversary or birthday greetings with the right number when you need it most.

A better letter. Personalize your notepaper or stick your initials to the back of an envelope to seal it. A decorative letter also adds style to a diary or school notebook which may then be covered with a protective plastic film.

Easy as 1-2-3. Have your door and mailbox numbers in a style you really like. By applying a coat or two of clear varnish to the numbers once mounted, you give them depth and permanence. A set of plain paper numbers is handy and fun for a party with a 'lucky draw' or sweepstakes.

Use your initialtive. Solo letters make smart ID tags for keys and such. Use heavy card and, after mounting, protect your tag with clear adhesive plastic coating for durability. Nothing to stop you making gloriously graphic earrings the same way.

A B C D E
F G H I J
K L M N O
P Q R S T
U V W X
Y Z

A B C D E
F G H I J
K L M N O
P Q R S T
U V W X
Y Z

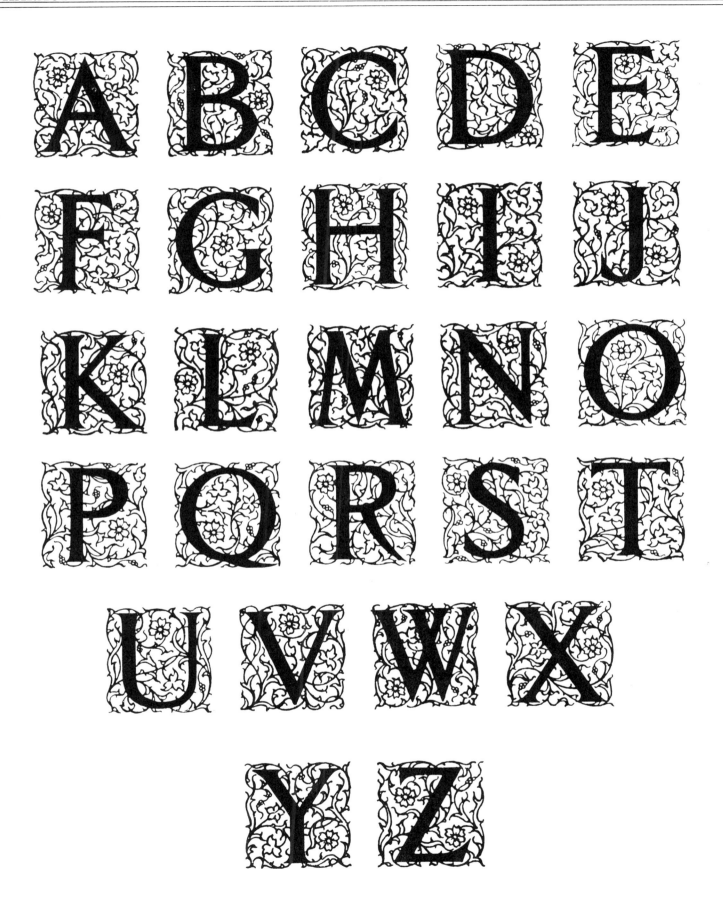

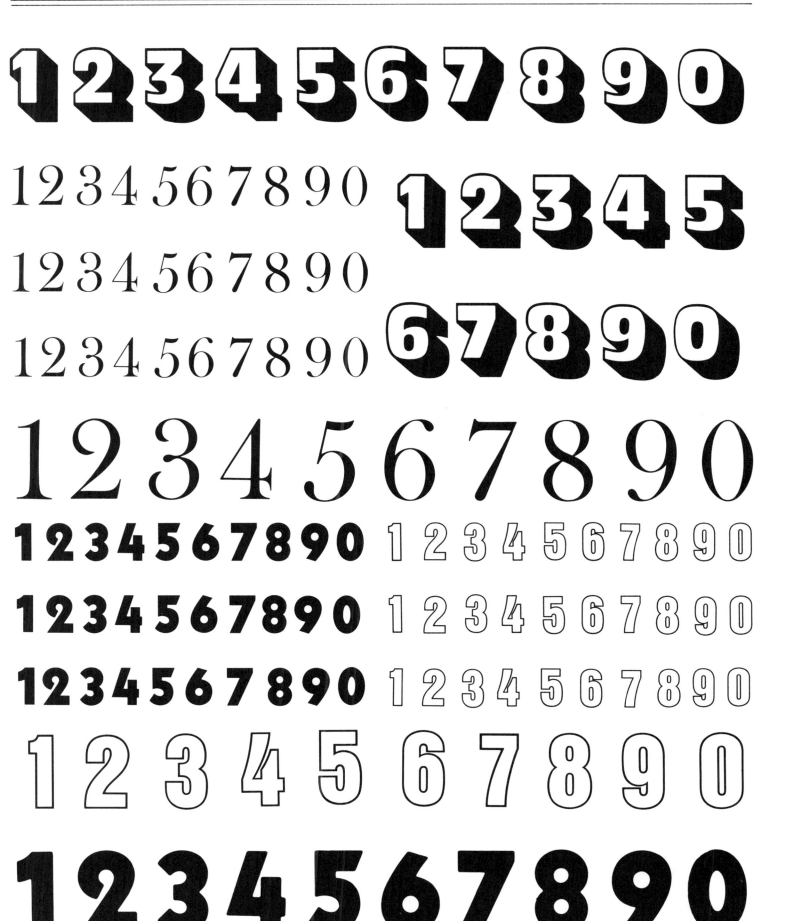

1234567890

1234567890   **1234567890**

1234567890   **1234567890**

1234567890   **1234567890**

1234567890

1234567890   **1234567890**

1234567890   **1234567890**

1234567890   **1234567890**

1234567890

1234567890

1234567890

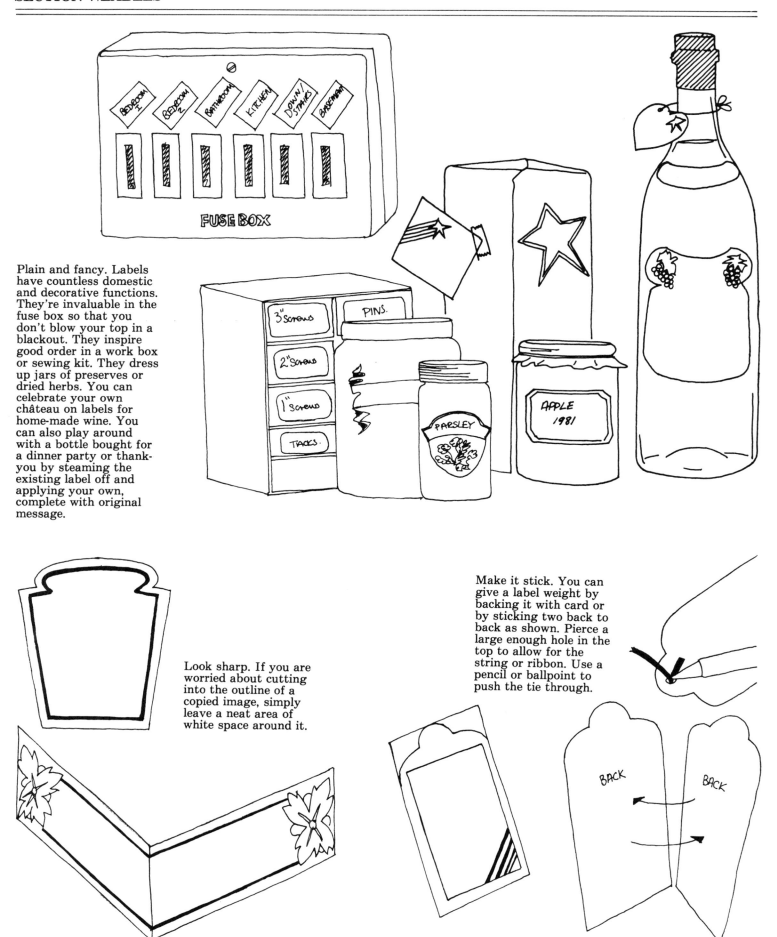

Plain and fancy. Labels have countless domestic and decorative functions. They're invaluable in the fuse box so that you don't blow your top in a blackout. They inspire good order in a work box or sewing kit. They dress up jars of preserves or dried herbs. You can celebrate your own château on labels for home-made wine. You can also play around with a bottle bought for a dinner party or thank-you by steaming the existing label off and applying your own, complete with original message.

Look sharp. If you are worried about cutting into the outline of a copied image, simply leave a neat area of white space around it.

Make it stick. You can give a label weight by backing it with card or by sticking two back to back as shown. Pierce a large enough hole in the top to allow for the string or ribbon. Use a pencil or ballpoint to push the tie through.

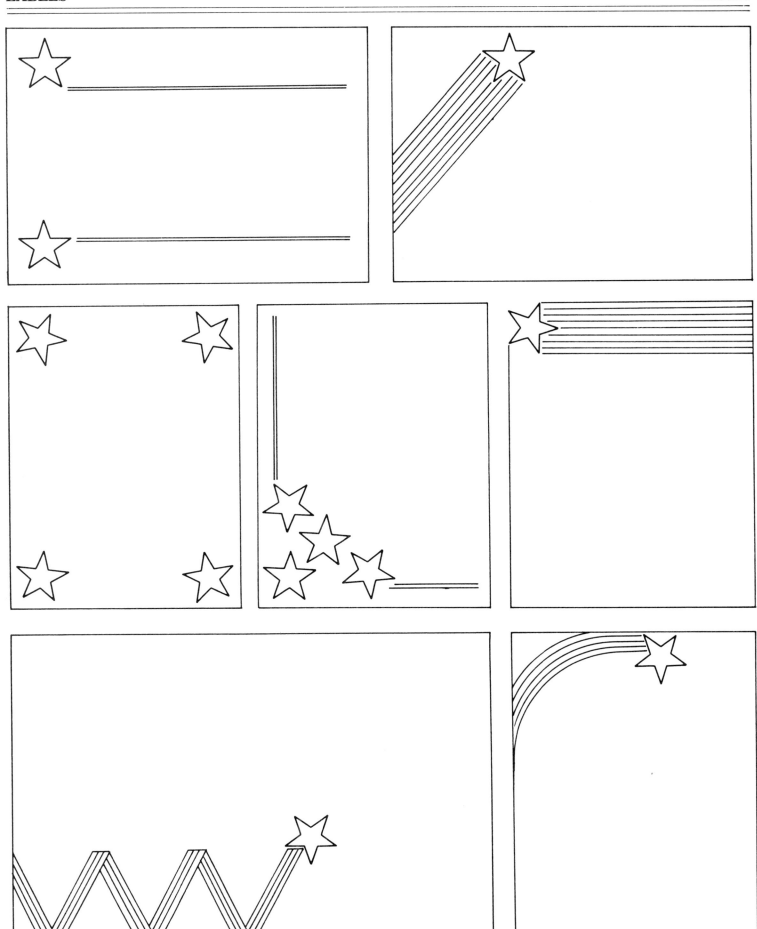

FROZEN ON

My cook book said this kinda stuff will survive .... months

I AM

I'll poison you if you don't defrost me by

I'm a herb by the name of

I'm ....
My ice-age began on

The Chef calls me

We made a date for no later than

See you around in...

Promise I can come out of the cold by...

I'll taste good anytime up to ....
By the way, my name is

Into the freezer

Must be out by

FROZEN ON

The good book said I only needed .... added
Plus a .... slow/fast cook
But you have to do it by ....

NAME

FROZEN ON

USE BY

Name

Frozen on

Use by

Name

Frozen on

Use by

Shopping list

Remember!

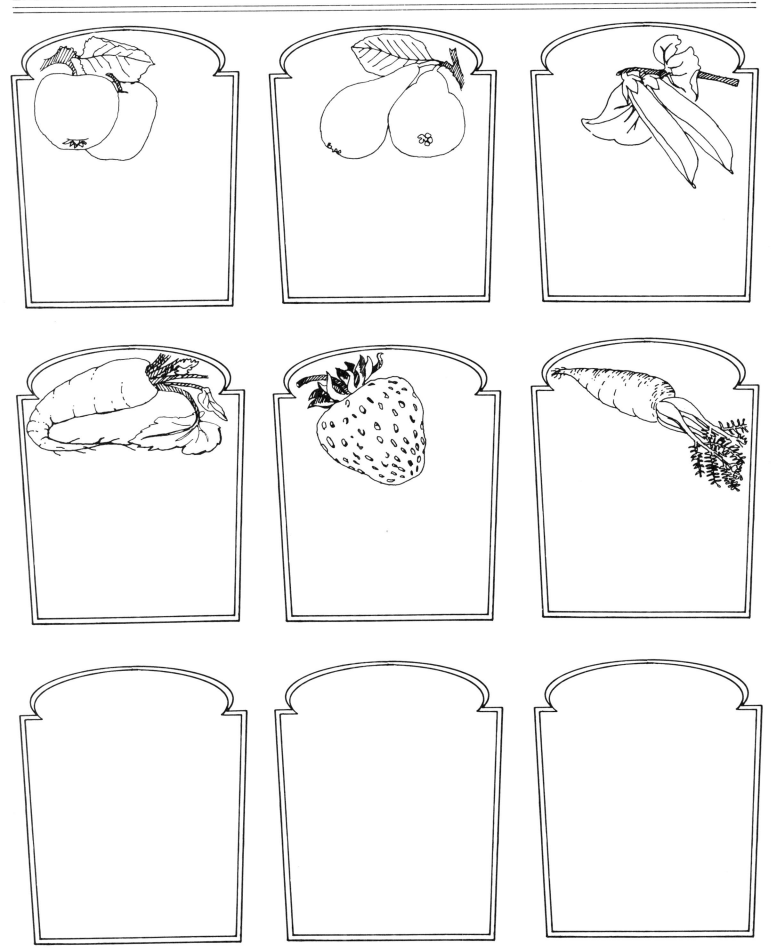

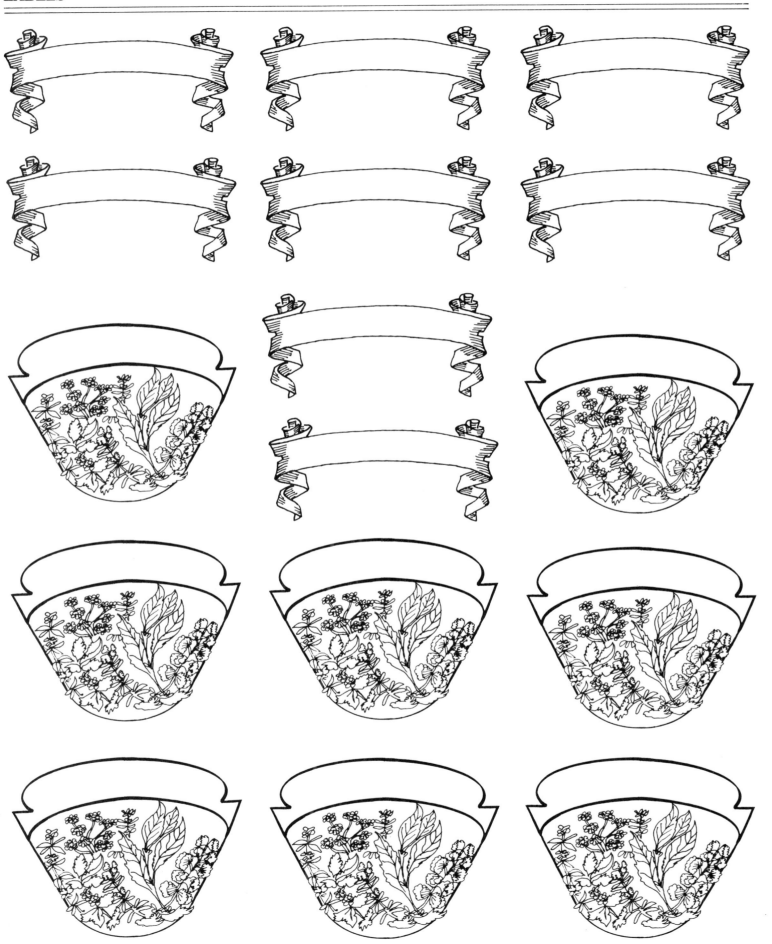

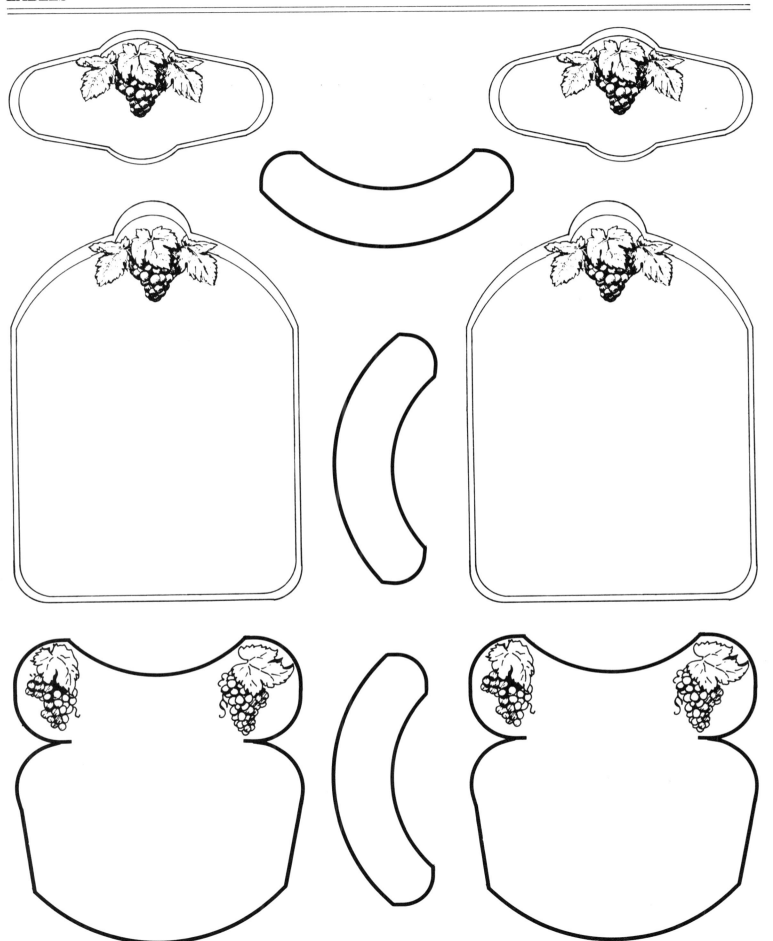

FOLD

FOLD

FOLD

FOLD

FOLD

FOLD

FOLD

FOLD

FOLD

FOLD

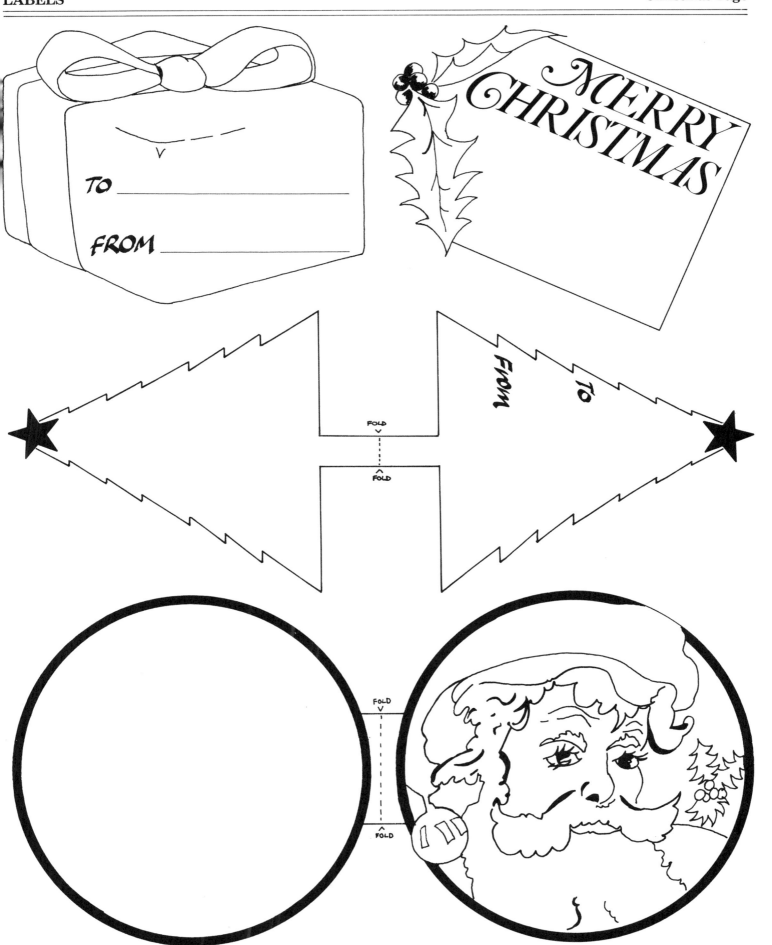

TO

FROM

MERRY CHRISTMAS

FROM

TO

FOLD

FOLD

FOLD

FOLD

A seductive slant. For a dinner party, clip an irresistible colour picture from a magazine. Mount it on white card and apply the 'Invitation' copy across one corner.

A solid offer. You can tie or paste your invitations for drinks or cocktails to swizzle sticks or disposable glasses and give your intended guests something to hang onto.

Inviting looks. Some of these invites are just begging to be coloured creatively so they'll have maximum appeal. Time flies as you ponder how best to draw in clock hands. Even grown-ups goof and make both the same size, so watch it.

The perfect host. Some people expect to have things handed to them on a plate, so indulge them with this three-dimensional invite. Maker a paper band to go around a paper plate, or, if you prefer, a brightly coloured paper napkin. For invites that mail easily, immortalize your own crockery by sticking the wanted words on a real plate and photocopying that.

An INVITATION

An Invitation

An Invitation

INVITATION

An Invitation

An INVITATION

Please join us for Dinner

DINNER
will be
served
at

YOU ARE
INVITED
TO
Dinner

YOU ARE
INVITED TO
Dinner

Please come to
DINNER

Buffet

BUFFET

Buffet

Buffet

BUFFET

BUFFET

Buffet

Buffet

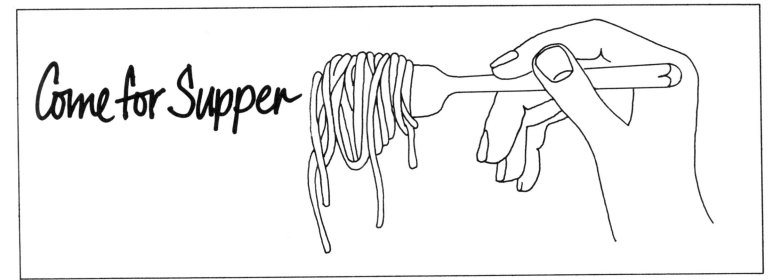

Come for Supper

SUPPER

SUPPERTIME
at

Barbecue

A
BARBECUE
at

BARBECUE

Barbecue

Barbecue

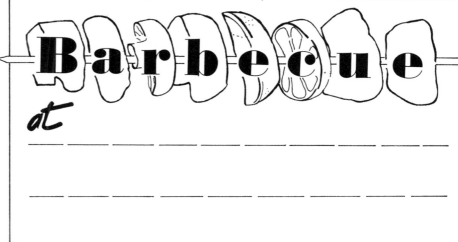

at

BRUNCH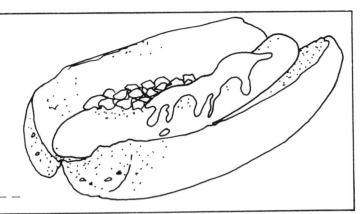

BRUNCH

Brunch

Brunch

Brunch

BRUNCH

COME BY FOR DRINKS

LOOK
IN FOR
A DRINK
AT

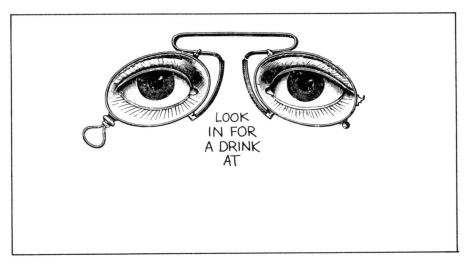

Time for a drink

AT

 *at our place for a drink*

DRINKS *will be served* *at*

You're welcome for **Drinks** *at*

DRINKS

Drinks

Cocktails

Drinks

COCKTAILS

COCKTAILS

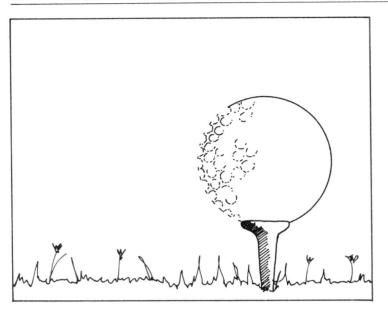

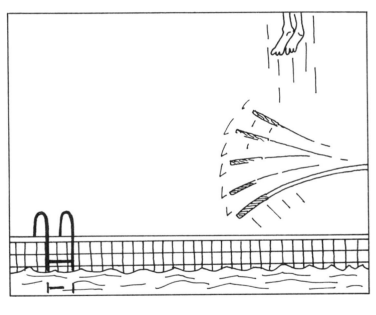

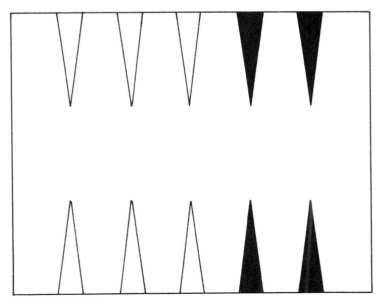

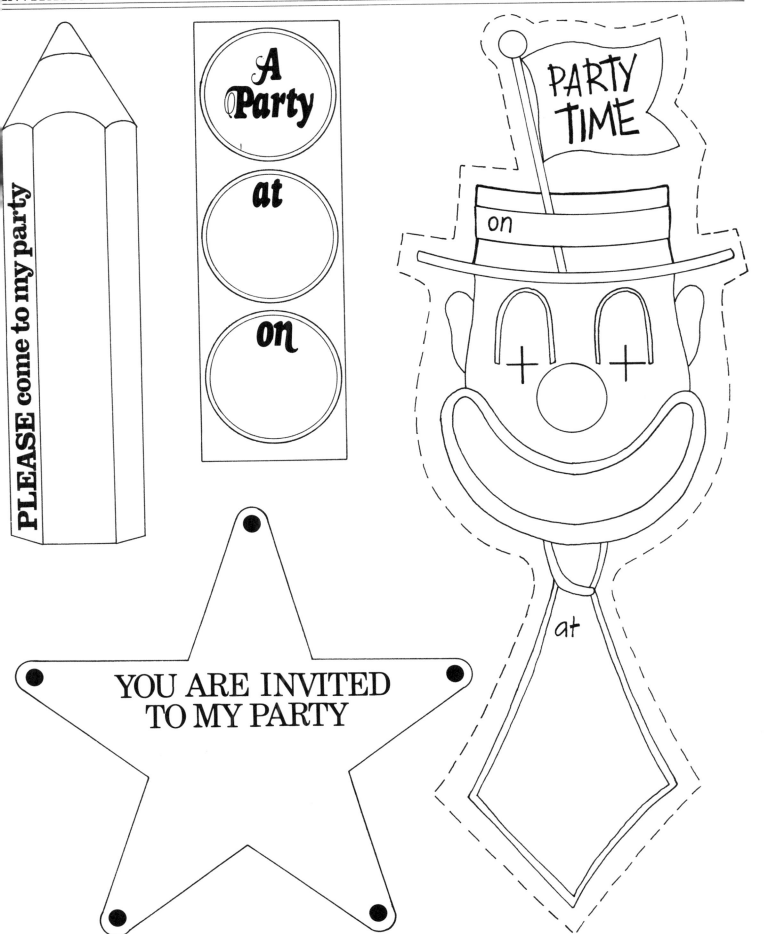

PLEASE come to my party

A Party

at

on

PARTY TIME

on

at

YOU ARE INVITED
TO MY PARTY

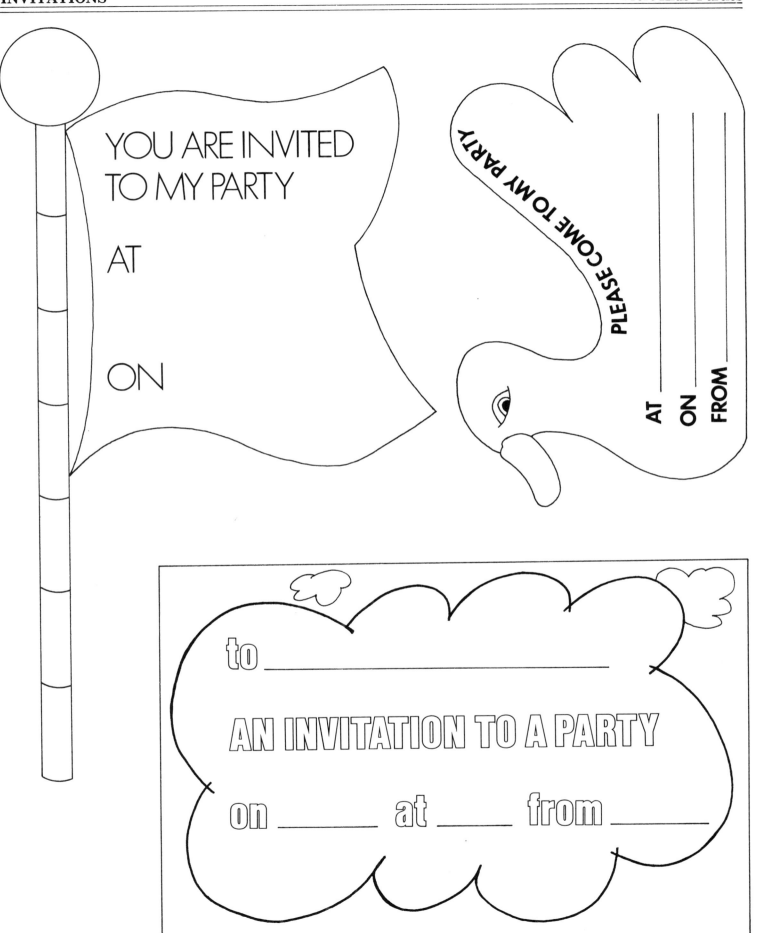

Please come to my party on...

from

A PARTY

at _____

on _____

You are invited to my party

Time _____

Place _____

From _____

To

Please come to my party

at

on

A party

Time

Place

R.S.V.P.

MY RECIPE FOR

INGREDIENTS

METHOD

Glad you enjoyed it

From the kitchen of

The celebrated recipe for

Ingredients

Method

Bon Appétit!

WHEN WE VISITED...

WE HAD A GREAT TIME AT...

Good contacts

Restaurants / Museums / Hotels
Stores / Bars / Cafes

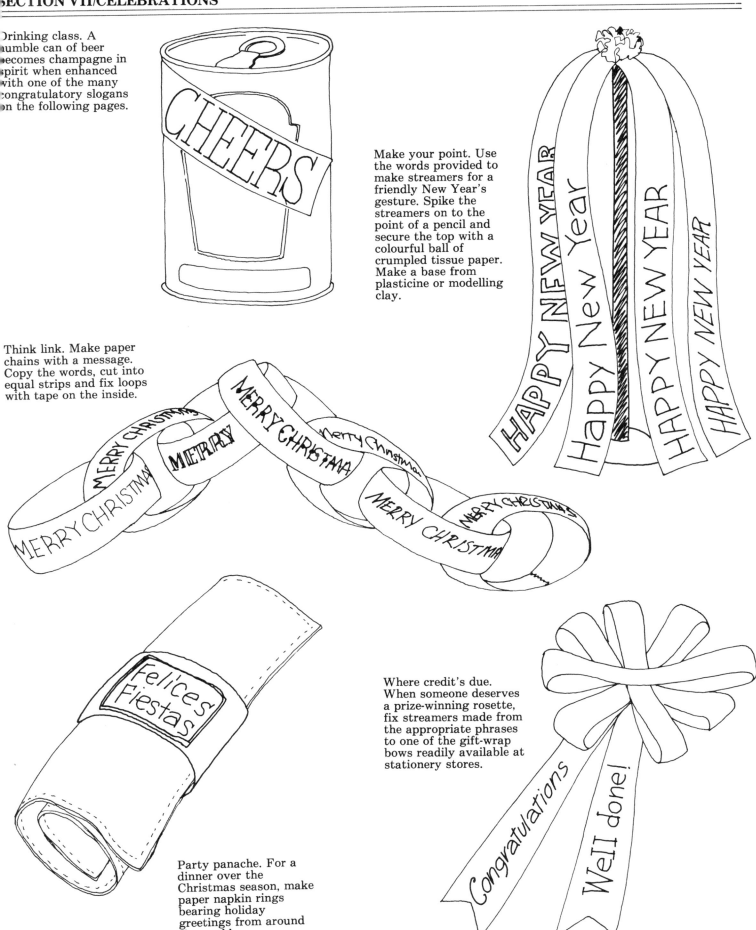

Drinking class. A humble can of beer becomes champagne in spirit when enhanced with one of the many congratulatory slogans on the following pages.

Make your point. Use the words provided to make streamers for a friendly New Year's gesture. Spike the streamers on to the point of a pencil and secure the top with a colourful ball of crumpled tissue paper. Make a base from plasticine or modelling clay.

Think link. Make paper chains with a message. Copy the words, cut into equal strips and fix loops with tape on the inside.

Where credit's due. When someone deserves a prize-winning rosette, fix streamers made from the appropriate phrases to one of the gift-wrap bows readily available at stationery stores.

Party panache. For a dinner over the Christmas season, make paper napkin rings bearing holiday greetings from around the world.

HAPPY BIRTHDAY

HAPPY BIRTHDAY

MERRY CHRISTMAS

Felices Fiestas

Meilleurs Voeux

Merry Christmas

HAPPY NEW YEAR

Happy New Year

Happy Chanukah

BONNE ANNÉE

HAPPY NEW YEAR

Merry Christmas

Season's Greetings

Happy New Year

Merry Christmas

Merry Christmas

Happy New Year

Name _____ Born at _____

Sex _____ Weight _____ Time _____

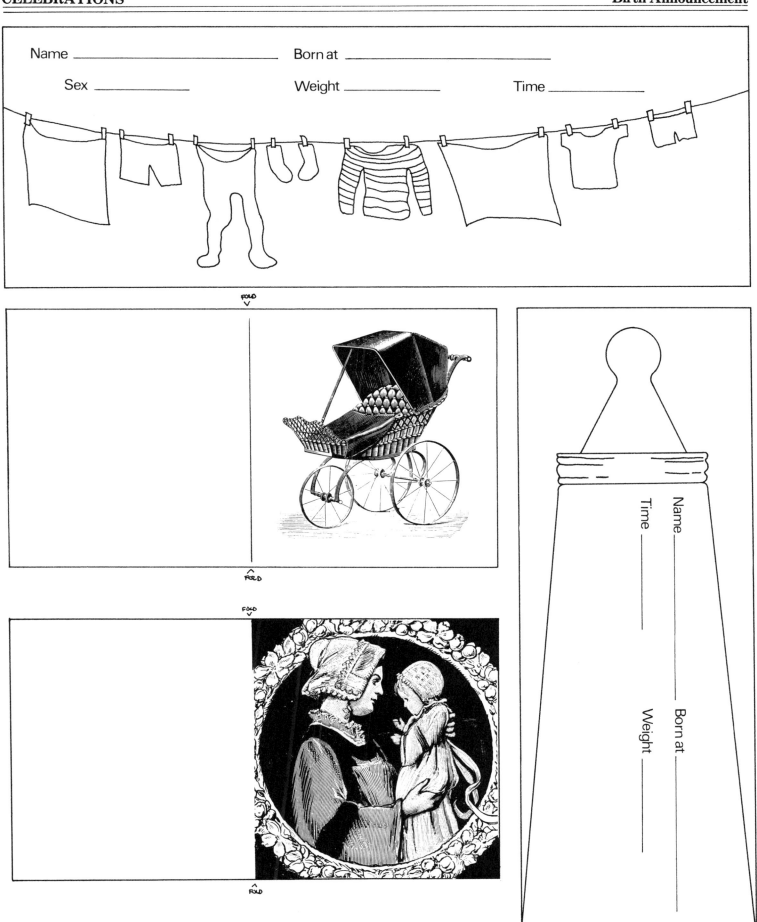

FOLD

FOLD

FOLD

FOLD

Name _____

Time _____

Born at _____

Weight _____

Congratulations

CONGRATS!!!

CONGRATULATIONS
_____
_____

Well done!

MAZEL TOV!

L'Chaim

CHEERS!

SKOL!

BIRTHDAY LIST

JANUARY

FEBRUARY

MARCH

APRIL

MAY

JUNE

JULY

AUGUST

SEPTEMBER

OCTOBER

NOVEMBER

DECEMBER

WE MUST WRITE
These 'THANK YOU'
Letters

TO
FOR
FROM

TO
FOR
FROM

TO
FOR
FROM

TO
FOR
FROM

TO
FOR
FROM

TO
FOR
FROM

TO
FOR
FROM

TO
FOR
FROM

TO
FOR
FROM

TO
FOR
FROM

TO
FOR
FROM

This Year 198___
we got Seasonal
greetings
from . . . .

All the trimmings. Cupboard doors and drawers are ideal surfaces for decorative borders. Colour the designs first. If you choose to use double-sided tape to apply them, put the tape on the back of the designs before cutting them out. Leave plenty of white around borders with fine lines as this helps set off the image.

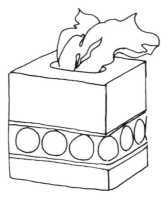

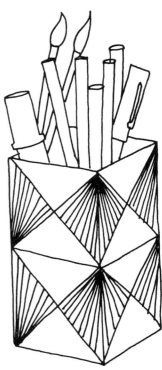

Boxing clever. Beautify a tissue box by using an attractive shade of paper as a background for hand-tinted borders. When making the cardboard frame for your pencil box, be certain the width of each side and the chosen design will match up.

Give someone the runaround. Commercial containers are often too good to throw away. Inventively decorated, they can be recycled as gifts filled with homemade goodies or kept for storage. Make a fresh start by covering the outside with spray paint. (Do this out of doors or in a well-ventilated room.) Fix your design around the perimeter with double-sided tape and apply a protective varnish as a finishing touch.

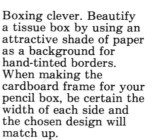

Small is beautiful. Little boxes can be decorated so that a simple present becomes an impressive presentation. Paste the pre-cut patterns into position and try a gift tag from the LABELS section for your message.

Give yourself the edge. Whether for invitations, announcements or personal notepaper, suitable border patterns make for improved appearance. You may think you deserve a star for writing that long-owed letter; stick it on the envelope and set a shining example for you correspondent.

Advanced booking. The motif of book spines may be repeated to create an attractive frieze along the top of a den wall, or it would be at home lining the wall behind bookshelves where it would naturally fill in the blanks. Fill in your own authors and titles as you like, including the bestseller you always meant to write. Remember, you can mask the tilted books if you want to produce a long row of the upright spines.

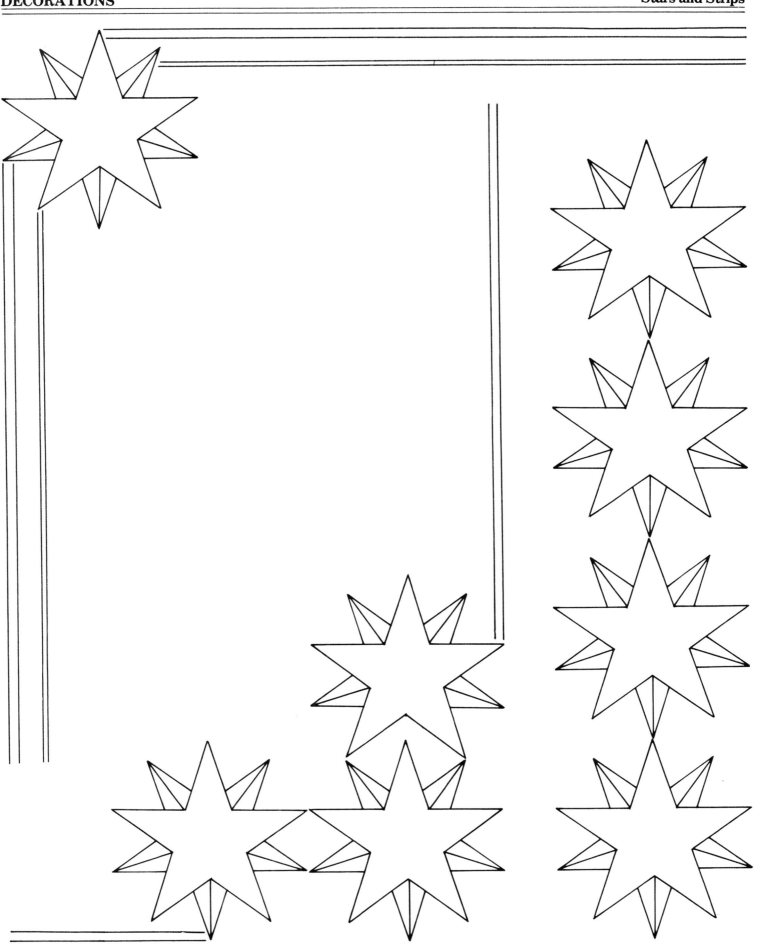

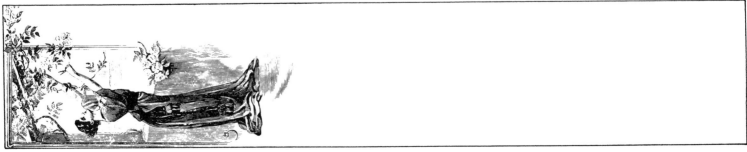

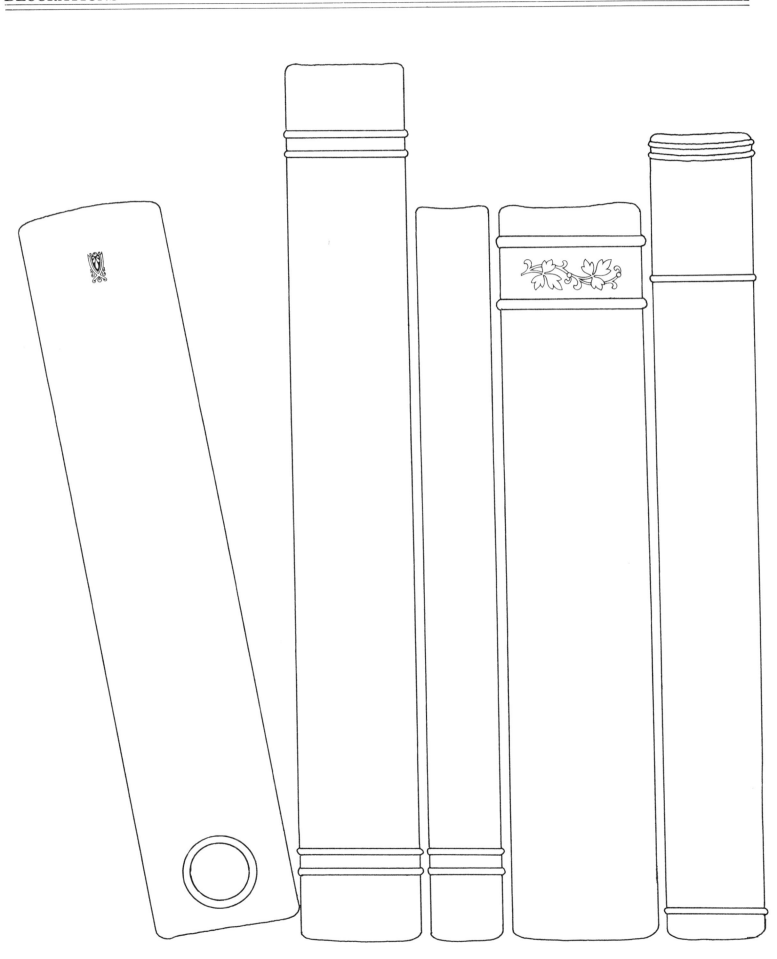

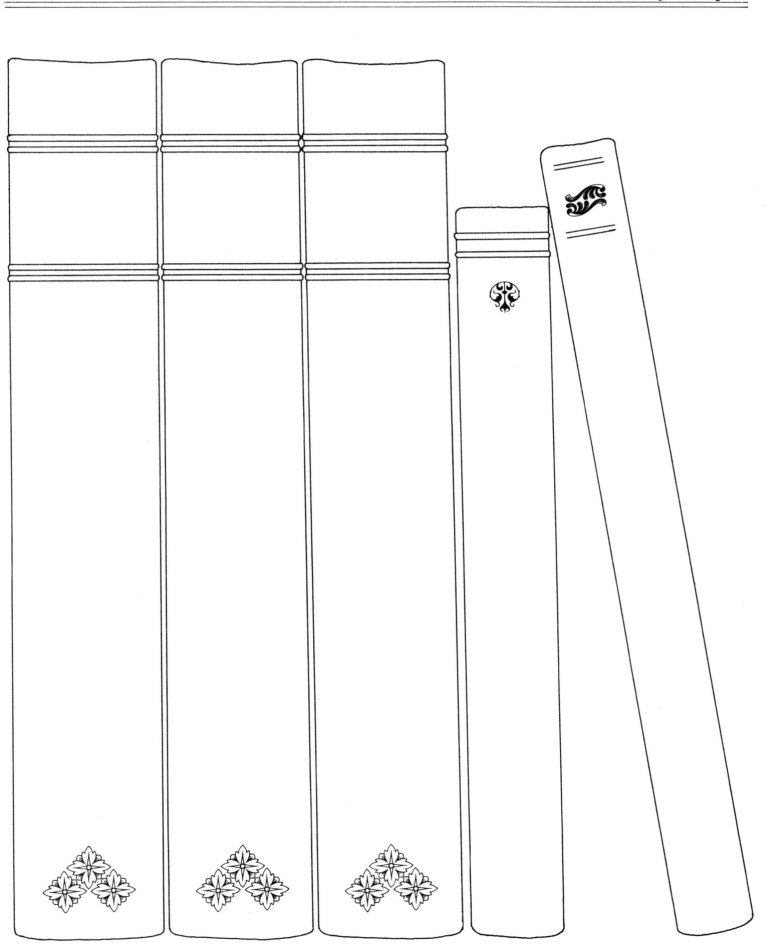

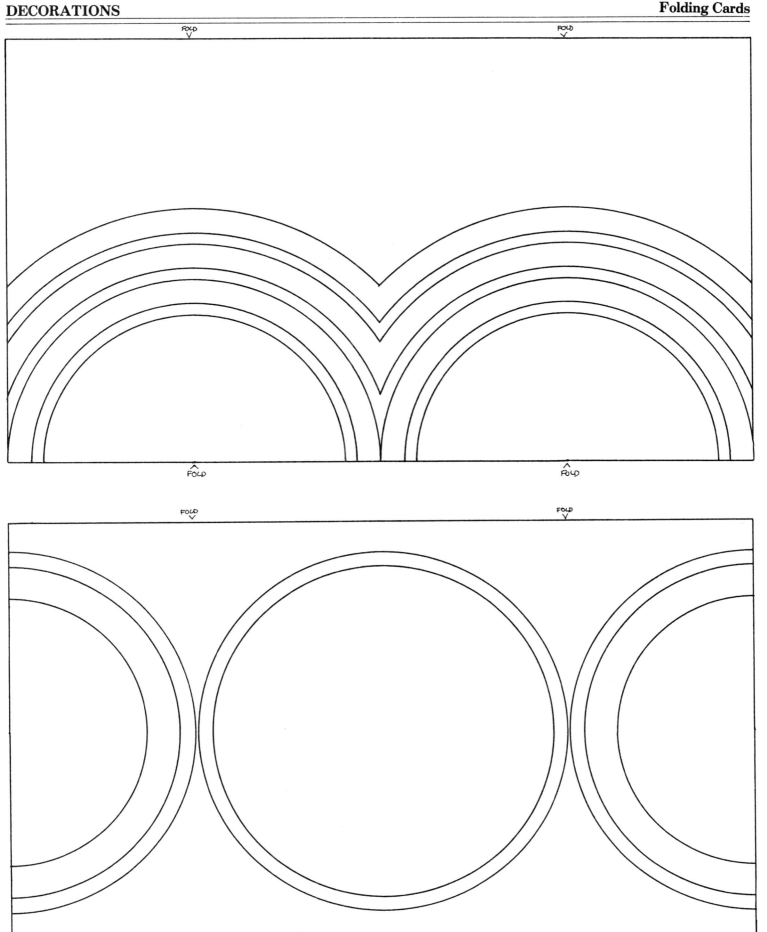

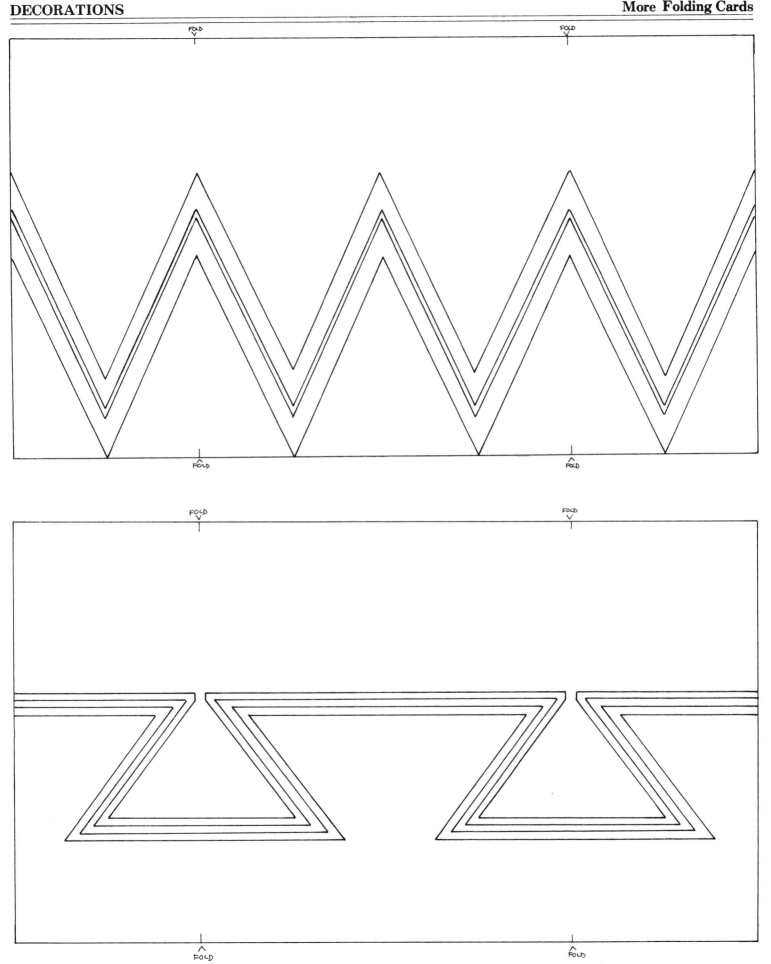

FOLD
FOLD
FOLD
FOLD
FOLD
FOLD
FOLD
FOLD